Mathematics

A review of inspection findings 1993/94

A report from the Office of Her Majesty's Chief Inspector of Schools

London: HMSO

ISBN 0 11 350050 5

Office for Standards in Education
Alexandra House
29–33 Kingsway
London WC2B 6SE

Telephone 0171-421 6800

Contents

Annex B

Annex C

Annex D

Introduction

This report is based on evidence from inspections of 79 primary schools by Her Majesty's Inspectors of Schools (HMI) to train Registered Inspectors (RgIs), and 735 secondary schools (including 45 middle schools deemed secondary) carried out by RgI teams using the OFSTED Framework. In most of the primary schools, all large JMI schools, mathematics was inspected by trainee RgIs. The analysis of lessons was based on 333 in Key Stage 1, 581 in Key Stage 2, 7280 in Key Stage 3, 4804 in Key Stage 4 and 1398 in the sixth form. Comments on the sixth form are based almost exclusively on A-level lessons; evidence is insufficient to comment on GCSE or other courses. Evidence from HMI survey work of mathematics in primary and secondary schools has been used to assist in the interpretation of patterns emerging from the analysis of the main database. In a few instances this has been used to plug gaps in material which was not forthcoming from institutional inspections.

The sections of this report concerned with inspection development have drawn on additional evidence; in particular that from the monitoring of inspections by HMI and the scrutiny of subject inspection evidence and mathematics sections in inspection reports.

The National Curriculum was fully in place in all years from Year 1 to Year 11; it was, however, being implemented for only the first time in Years 6 and 11. In 1994 the statutory assessment arrangements were in place in Key Stages 1 and 3 while at Key Stage 2 tasks and tests were being piloted on a voluntary basis. Secondary schools were preparing for new GCSE syllabuses in 1994 based on the National Curriculum. Changes in the reporting system – that is, a move away from awarding in relation to National Curriculum levels to a grade system based on marks – were announced after teachers had started to teach the GCSE course.

Subject Report

Main Findings

- There are concerns about **standards of achievement in relation to pupils' capabilities** in about a third of the primary schools and a fifth of the secondary schools. The standards are satisfactory or better in three-quarters of the schools at Key Stage 1, in two-thirds at Key Stage 2 and in four-fifths of the secondary schools. (Paragraphs 1, 3)

- The primary schools give a lot of attention to routine number work and standards in Key Stage 1 are often good and sound overall. However, in Key Stage 2 standards are less satisfactory mainly because the rate of progress is too slow and misconceptions and errors are not addressed. Too many pupils are unable to recall important number facts or to compute with sufficient speed and accuracy. (Paragraph 1)

- In Key Stages 3 and 4 standards are variable and, in almost a quarter of the schools, standards of numeracy are too low. Although most schools ensure that pupils are competent with basic number facts and skills many do not pay sufficient attention to keeping these skills sharpened. (Paragraphs 4, 6)

- Almost all the schools have made satisfactory **progress in implementing the National Curriculum** but insufficient attention is given in many schools to teaching the processes and skills associated with using and applying mathematics (AT1). Handling data (AT5) is still under-developed even though most teachers recognise this as an important aspect of the mathematics needed for *everyday life*. (Paragraphs 32, 33)

- **Challenge, pace and motivation** in lessons are significant factors in pupils achieving good standards. Although **attitudes to learning the subject are positive** in most schools, too many lessons, particularly in Key Stages 2 and 3, lack **challenge and pace**. (Paragraphs 15, 22)

- Greater awareness of achievements in primary and middle schools has led to a better match of work on transfer to secondary school. (Paragraph 16)

- The schools which achieve consistently good standards are invariably **well managed by subject co-ordinators or heads of department** and clear guidance is given to teachers. Few primary co-ordinators have time to develop their role. (Paragraph 42, 43)

- Nearly all the schools have devised systems for assessing and recording the achievements of pupils which **meet statutory requirements**. However, few schools have written policies to guide teachers' planning for assessment and to ensure that the results are appropriately used to inform future teaching, ensure continuity and progression and promote higher standards. (Paragraph 26)

Key issues for schools

The following issues are common to primary and secondary schools. Schools need to consider:

- ways of raising standards, particularly in numeracy;

- if the teaching provides sufficient challenge to pupils of all abilities, particularly in Key Stages 2 and 3;

- how they make more effective use of the time for mathematics;

- the nature and quality of the guidance that is given to teachers;

- the effectiveness of assessment, recording and reporting policies and practice, particularly in relation to planning future teaching programmes;

- how they monitor the quality of work in classrooms and the match of work to the pupils' previous attainment;

- how they provide progression as pupils move within schools, and also when they change schools;

- how they make more effective use of the resources they have as well as evaluating whether they have a sufficient range to support the curriculum;

- how to fully meet statutory requirements.

Primary schools need to consider:

- how to raise expectations in Key Stage 2 especially and support teachers who lack confidence;

- how mathematics co-ordinators can be given sufficient non-teaching time to carry out their role.

Secondary schools need to consider:

- how they help pupils to keep sharp skills acquired in the earlier years, and others developed later, as they progress through the compulsory years of schooling.

Standards of achievement

The GCSE and GCE results achieved nationally in mathematics in 1994 are addressed in a section commencing at paragraph 65.

Key Stages 1 and 2

1   **Standards of achievement in terms of pupils' capabilities** are at least satisfactory in three-quarters of the schools at Key Stage 1 and in two-thirds at Key Stage 2. Consequently there are concerns about standards in a third of primary schools. In both key stages, more schools have satisfactory standards in knowledge and skills than understanding. In many lessons, pupils put most energy into practising their skills and consolidating their knowledge; they are much less competent at explaining their work and applying their knowledge (AT1). Standards in AT1 are often unsatisfactory, especially those of able pupils. In both key stages, work in number (AT2) is given a lot of attention. At Key Stage 1 **standards of achievement in number** are often good and sound overall with the majority of pupils developing a good knowledge of basic skills, including mental work. The picture is more variable in Key Stage 2 where standards are less satisfactory mainly because the rate of progress is too slow and misconceptions are not addressed. Too many pupils are unable to recall important number facts or to compute with sufficient speed and accuracy.

2   Standards in handling data (AT5) range widely between schools from good to unsatisfactory. Pupils in Key Stage 1 sometimes produce good diagrams and graphs linking two variables, for example eye colour and names. In Key Stage 2, pupils are reasonably confident with the early ideas of probability but few pupils are able to refine their ideas and use language more rigorously. Standards in algebra (AT3) and shape and space (AT4) also vary considerably. At Key Stage 2 initial work with coordinates is understood well. However, it is rare to find pupils extending this knowledge and understanding beyond routine and sometimes tediously repetitive examples. Pupils' knowledge and understanding of the volume and nets of simple three-dimensional shapes is particularly good where they have had early tactile experience of a range of solid shapes. In too many schools insufficient attention is given to the precise use of mathematical language and appropriate levels of accuracy.

Key Stages 3 and 4

3   Pupils' **standards of achievement in relation to their capabilities** are at least satisfactory in four-fifths of the schools, particularly in relation to knowledge and skills. Where standards are good pupils have opportunities to air and discuss their conceptions and misconceptions; as a result of feedback from teachers or other pupils, their understanding is more secure. This understanding enables the pupils to apply their knowledge and skills accurately to a range of problems across all the attainment targets. For example, they are able to find solutions in number work when they have forgotten a standard method, draw graphs accurately and interpret them sensibly, process data and appreciate the significance of the different measures of average.

4   **Standards in numeracy** are slightly lower than those in mathematics. In almost a quarter of the schools standards in numeracy are too low. Pupils are a little better at applying their mathematical knowledge and skills in Key Stage 4 than Key Stage 3 and are more successful in applying them in mathematics lessons than in other subjects. Frequently opportunities for pupils to use their mathematical skills in other subjects are not exploited by the teachers mainly because the schools rarely plan ways of developing numeracy across the curriculum.

5   Where standards of numeracy are good, pupils successfully apply their skills in subjects such as science, geography and technology. These pupils measure and calculate accurately, draw and interpret graphs, record and present data appropriately and use their knowledge of shape and space in making and evaluating artefacts. Good standards occur where pupils are competent in mental mathematics, confidently discuss applications of number in a variety of situations across the curriculum and use calculators effectively. Pupils in these schools are successfully learning of the power of mathematics to predict and explain.

6   Standards of numeracy are poor where demands across subjects are ill-matched to the pupils' competencies and previous experiences. For some pupils, a poor recall of number facts and a lack of mental agility with number results in inaccurate work and a lack of

confidence. Almost all the secondary schools ensure in Year 7 that pupils are competent with basic number facts and skills but many do not pay sufficient attention to keeping these skills sharpened.

7   Standards in applying mathematical skills and knowledge (AT1) are less well developed than the other attainment targets although in some schools there are concerns about standards in algebraic manipulation. In a few instances, shortages of resources and long queues of pupils waiting for help result in some under-achievement. In other classes, pupils have no time or encouragement to think about their mathematics which, when combined with poor monitoring by the teacher, means that some pupils remain confused at the end of a topic, for example, about the difference between perimeter and area or multiplying decimal fractions by 10.

Sixth Form

8   Standards achieved by sixth form students (relative to their capabilities) on A-level and AS courses are at least satisfactory in nine-tenths of the schools and are high or very high in almost a quarter. Most students are coping well with the level of work expected of them and are learning to apply new and more rigorous approaches to their study of mathematics. Standards are particularly high when students are at ease with algebraic manipulation and are able to demonstrate a high level of critical and analytical thinking.

Quality of teaching

Quality of teaching by Key Stage

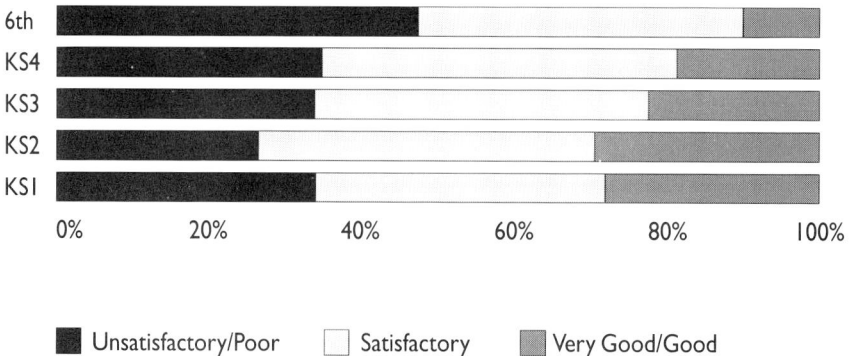

Legend: Unsatisfactory/Poor   Satisfactory   Very Good/Good

Key Stages 1 and 2

9   In primary schools, the **quality of teaching** is generally better in Key Stage 1 than Key Stage 2. It is satisfactory or better in 72% of the lessons at Key Stage 1 and 70% of the Key Stage 2 lessons; it is good or very good in 35% of the lessons at Key Stage 1 and 27% of the lessons at Key Stage 2. The relatively small proportion of good or very good lessons in Key Stage 2 and the higher proportion of unsatisfactory and poor lessons compared with other key stages are a matter of concern.

10   The **objectives for lessons** are clear in about two-fifths of the schools. Where objectives for learning are very clear, there is a plainly-stated purpose with well-defined tasks for individual pupils. The lack of clear objectives is closely linked to poor planning, both at individual teacher and school level, weak co-ordination of the subject through-out the school and a shortage of guidance for teachers.

11   The teachers' **command of mathematics** is adequate for the work they are called on to teach in nine-tenths of the schools at Key Stage 1 and in three-quarters at Key Stage 2 but many lack confidence. In only one school in ten, however, is the teachers' command of the subject good or very good. More Key Stage 1 teachers have experience of the National Curriculum, and clearly-focused INSET linked to National Curriculum requirements and improved knowledge of progression within attainment targets are factors which contribute to the difference between the key stages. The subject knowledge of those teaching able pupils at the end of Key Stage 2 is generally satisfactory but this is because the emphasis is frequently on number and the teachers feel secure and confident. Areas of weakness in subject knowledge are clearly revealed when teachers cannot extend pupils' learning by exploiting opportunities which arise, for example, where work on symmetry remains a routine textbook exercise even where pupils are bringing leaves to class as part of a science study. Too many teachers, especially in Key Stage 2, lack an understanding of the way pupils learn, or do not learn, mathematics so are not aware how to help them overcome their mistakes.

12   There are very significant differences in the quality of the content of the lessons and the activities chosen to meet the aims and objectives in each key stage. The **lesson content and activities** are broadly suitable

for their purpose in over four-fifths of the schools at Key Stage 1, but only in seven-tenths of the schools at Key Stage 2. These differences are significant and are indicative of the fact that too many pupils in Key Stage 2 spend a lot of time working individually through texts, with little direction from, or interaction with, the teacher. This frequently results in a mismatch of work in terms of pupils' previous achievements and no progress in learning taking place.

13   Pupils are almost all taught mathematics by their class teacher and for a significant proportion of the week. Usually ability groups are formed within the class. Whole class sessions, even short ones, are rare. Some short, effective class sessions were observed where pupils rehearsed previous knowledge of numbers or shapes, or discussed ways of carrying out a survey. Too often time is wasted when pupils set the pace of the work, or when they colour in patterns or shapes to no purpose.

14   In a large proportion of lessons teachers provide practical or written activities which are broadly suitable for children in the middle ranges of ability for the classes taught. In many cases pupils are organised in small groups within classes; the content or activity for each group, however, is not always well matched to individual capabilities. In particular at both key stages too many tasks are undemanding for the ablest children present whose achievement is thereby restricted. Many teachers circulate between groups well, often offering sympathetic support to the lowest attainers and responding to requests for further information from all children. A weakness of many lessons is that the teacher's time is spent mainly with those who are stuck and insufficiently in direct teaching with all pupils. Too many able pupils, or the conscientious quiet ones, do not have interactions with their teachers. In the most successful classes a good range of organisational strategies is used, resources are utilised effectively, tasks and content are varied to suit pupils' individual needs, there is regular feedback to pupils about both accuracy and methods adopted. Teachers are frequently assisted by non-teaching staff or voluntary helpers. Where joint planning occurs and guidance is good, the additional adults make a significant contribution.

15   Most pupils respond willingly to tasks set, show great levels of commitment and concentration and generally enjoy mathematics. In

Key Stage 2 this is frequently in spite of dull and uninspiring teaching. **Challenge, pace and motivation** of lessons are the weakest features of much of the teaching, especially in Key Stage 2. Teaching that motivates pupils, providing constant challenge and a lively pace, is the most effective factor in the achievement of high standards. These features are broadly acceptable in about four-fifths of the schools in Key Stage 1 but in only three-fifths of the schools at Key Stage 2. In the good lessons, pupils of all abilities are appropriately challenged by tasks well matched to their abilities, interest is sustained with opportunities to work individually and collaboratively, and pupils work at a good pace. In the better lessons teachers praise pupils' achievements and questioning is perceptive and challenging.

Key Stages 3 and 4

16   The **quality of teaching** is of concern in about a fifth of the schools. It is slightly better in Key Stage 4 than Key Stage 3. It is good or very good in 36% of the lessons observed in each Key Stage; it is satisfactory or better in 78% of the lessons at Key Stage 3 and 81% at Key Stage 4. It is usually better with upper ability classes. In all years, but particularly in Years 7 and 8, the teaching is less than satisfactory more frequently with classes of middle attaining pupils. Greater awareness of achievements in primary and middle schools has led to a better match of work on transfer to secondary school.

17   In about two-thirds of the schools, the **objectives for lessons** are clear. However, in approximately one school in seven at Key Stage 3, lessons are not adequately planned, compared with a tenth of the schools for Key Stage 4. In the best schools new targets are set for each lesson and planning is based on good records. Without clear objectives work is the poorer for a variety of reasons. Insufficient work is securely based to progress from that attempted earlier, particularly in Year 7. But the greatest weakness in the planning and therefore the execution of lessons is that individual needs are inadequately considered. Too often work is not well matched to the previous attainments of pupils. This frequently results in a lack of sufficient challenge for the ablest pupils present, especially where the range of ability in a class is wide. However, it also occasionally results in the supply of an inappropriate range of reading material for poor readers.

18   Teachers have a good or very good **command of the subject** in three-fifths of the schools; it is adequate in almost all but slightly worse in Key Stage 3 where there is a higher proportion of non-specialist teaching. This is occasionally offset by recent and relevant in-service training. A good knowledge and overview of the subject allows teachers to plan work well both in the short and the longer term. In particular, it enables some teachers to insist on and develop a sound basis of mathematical language.

19   The lesson **content and activities** are broadly suitable for the purpose in both key stages in four-fifths of the schools. In the best schools, teaching approaches are appropriately varied from lesson to lesson, resources are used imaginatively, pupils improve from good feedback about previous work and exposition is very clear. In weak schools, the range of teaching approaches is often narrow, such as where there is much exposition by the teacher but very little opportunity for pupils to participate and respond, or where pupils are expected to learn too much on their own. In many weaker lessons, pupils spend long periods following set routines but do not learn when and how to use them.

20   In a very high proportion of schools, AT1 tasks are attempted at set times in most or all year groups, but are rarely closely related to other work. Frequently, these tasks are for pupils to investigate patterns in number or shape; in many schools insufficient attention is given to the range of these tasks and to progression within them. In a small but growing number of schools, work for AT1 is well integrated with other mathematical content; in the majority of these schools work is carefully planned for progression in the three strands of application, mathematical communication, and reasoning, logic and proof. In the majority of schools more attention needs to be given to the teaching of all such strands rather than using AT1 tasks only for assessment.

21   It is rare to find computers offering regular support or being used routinely in the mathematics classroom. In the best practice, information technology (IT) is used to teach a range of difficult mathematical relationships and applications. In many schools pupils are introduced to a few set tasks such as the use of spreadsheets or databases, or the control of a screen turtle to demonstrate some geometrical properties of two-dimensional shapes, for short periods.

22   In nine-tenths of the schools pupils' attitudes to the subject are positive. The majority of pupils are motivated to succeed and persevered at the tasks set. As at Key Stages 1 and 2, teaching that motivates pupils, providing constant challenge and a lively pace, is the most effective factor in the achievement of high standards. The **challenge, pace and motivation** provided by the teaching is ineffective in three-tenths of the schools at Key Stage 3 and one-quarter at Key Stage 4. In the best schools, expectation is high, pupils' interests are sustained, and questioning is of a high quality within a lively presentation. In weak lessons teachers are unaware of pupils' lack of understanding, and questioning does little to develop concepts. Pupils set the pace and waste time particularly when they begin to lose their concentration in long lessons, or wait for long periods for help from the teacher. In good lessons teachers determine the pace and plan how to use the time available. A feature of many good lessons, observed in Key Stage 4 particularly, was a closing whole class session in which the teacher drew together the main points of the lesson effectively, drawing out conclusions from pupils. Good sessions were observed with younger pupils who worked mainly individually but were drawn together at the end for mental work or to allow a group of pupils to explain to others what they had been doing. Too many individualised lessons drift to a close but here time was used productively.

Sixth Form

23   The quality of teaching is satisfactory or better in 89% of A-level lessons and good or very good in 47%. Teachers generally have a good command of their subject and are well prepared and organised with clear objectives for lessons.

24   In the best lessons, teachers make good use of a range of teaching strategies: clear exposition linking new knowledge and skills with prior learning, discussion, practical work with computers and targeted questioning which attempts to involve all students. In these lessons teachers are responsive to students' needs and give timely support to individuals. Through well chosen interventions teachers provide a balance between giving students new ideas and challenging them to reach their own conclusions.

25   In some lessons teachers lack confidence and imagination,

preferring to follow procedures mechanically. In these lessons teachers tend to attempt uncompromising progress through a textbook and make little attempt to involve students in discussing their work. Students are inadequately challenged to think for themselves and find it difficult to maintain their interest.

Assessment, recording and reporting

26   Nearly all schools have developed systems for assessing and recording the achievements of pupils which **meet statutory requirements**. Effective systems contribute to the achievement of good standards but, in reality, the effectiveness varies widely. Few schools have devised policies which make clear how the data collected on pupils' attainments could guide teachers in modifying their future planning for individuals and classes and, in the longer term, to ensure continuity and progression.

27   Most teachers have a good general knowledge of the pupils' strengths and weaknesses and give appropriate oral advice on how to overcome short-term problems. However, the marking of written work is seldom used to diagnose specific conceptual problems or to inform teachers' planning for subsequent work. In many secondary schools this is exacerbated by the fact that **marking** is frequently undertaken by the pupils themselves. Too often this practice merely indicates inaccurate answers and, unless closely monitored by teachers, does little to diagnose misconceptions or the use of inappropriate methods. Teachers' written comments are usually of a general nature and are rarely used to guide pupils' further study.

28   Teachers use a range of tasks in their day-to-day assessments of pupils' progress but much of the information so gained is not used to contribute to pupils' records of attainments. Formal assessments in Key Stages 3 and 4, and increasingly so in Key Stage 2, rely heavily on tests incorporated into published schemes or other written tests. These tests are frequently narrowly focused and fail to assess pupils' abilities to bring together different aspects of their mathematical knowledge and apply it in new and unfamiliar situations. Teachers rarely build assessment opportunities into their planning of classroom tasks.

29   The majority of recording systems indicate the curriculum that has been covered rather than the knowledge, skills and understanding that have been acquired. Most schools meet the statutory requirements for reporting to parents although anticipated changes in National Curriculum requirements have delayed the development of reporting systems to provide more detailed information in relation to specific attainment targets.

30   In Key Stages 3 and 4 AT1 (using and applying) is most often taught and assessed through extended pieces of classwork and homework. A minority of schools have developed agreed criteria for assessing this work but internal moderation of teachers' judgements is largely restricted to GCSE coursework. There is a widespread need for the development of procedures to ensure teachers are consistent in their judgements.

31   In the sixth form, work observed is closely matched to external examination requirements. Students' work is generally marked conscientiously and teachers give regular feedback to the students. The recent introduction of a number of new A-level courses, and the growth of modular syllabuses, have resulted in significant changes in methods of assessment for some students. Coursework and examinations spaced throughout the year are spreading the assessment load. Staff and students are enthusiastic about these new styles of assessment, which are having a positive effect on the range and quality of A-level work as well as the motivation of students.

Curriculum content

32   **Statutory requirements are being implemented** satisfactorily in nine-tenths of the schools. Most of the schools are aware of areas which need development. In many of the primary schools insufficient attention is given to AT3 (algebra) and AT5 (handling data). In many secondary schools handling data (AT5) is still under-developed even though most teachers recognise this as an important aspect of the mathematics needed for *everyday life*. Weakness often occurs where schools rely too heavily on a single textbook scheme which has not been updated to match National Curriculum requirements.

33   Limited progress is being made in planning for the implementation of AT1 (using and applying) and the use of IT to aid the teaching and learning of mathematics. However, the extent to which schools are **implementing AT1** (using and applying) into their teaching and learning programmes is unacceptably varied. In a significant proportion of schools, work in AT1 is not planned at all, and very little occurs incidentally.

34   Good **curriculum planning and organisation** are very significant factors in the achievement of high standards. The overall planning and organisation of the curriculum to ensure consistency of approach and high expectations is unsatisfactory in a fifth of the secondary and middle schools and in half of the primary schools. Some of the primary schools have their own broad policy statements which set out recommended approaches to teaching and learning; few are developing these statements to provide schemes of work, and not all teachers follow the recommendations. The implementation of such recommendations is rarely monitored and in practice commercial schemes continue to provide the basis for the mathematics curriculum in the majority of schools.

35   The majority of mathematics departments have some written **policies** and these are often contained within an overall **scheme of work** or a departmental handbook. Many schemes of work have been recently revised and contain guidance for all teachers on planning both of overall syllabuses and of day-to-day work. Effective schemes offer advice about appropriate teaching strategies and the use of resources, with suitable variations for different classes. A few schemes of work which are dated or have been hastily constructed do not reflect classroom practice. Where there is insufficient detail to guide the teachers, pupils sometimes work at inappropriate levels or receive unbalanced programmes. Where new A-level courses are being introduced, teachers have planned together more than in the past, and this has led to a greater variety of teaching styles and more effective learning.

36   **The amount of time allocated** to the subject is similar to previous years, that is about a fifth of taught time in primary schools and, most commonly, 175 to 185 minutes per week in secondary schools though there is an overall spread from 140 to 250 minutes per week. The effective use of the time available is a significant factor in achieving high standards.

37   In the primary schools, pupils are almost all taught mathematics by their class teacher and usually ability groups are formed within the class. The **organisation of classes** in secondary schools varies, but at some stage, in virtually all the schools, pupils are placed in sets according to their ability in the subject. Occasionally this is from the time of their arrival in secondary school; where this is the case the groups formed do not always take sufficient note of the pupils' primary school experience. A few good examples were found which show the pupils that the secondary teachers value the work they have done previously. For example, in one school where liaison is good, pupils arriving in Year 7 are greeted by a display of their work completed at various contributory primary schools in the previous term. Setting most frequently takes place at the end of Year 7 or 8. Where practice is good the placement of pupils is based both on teachers' own assessments and on the evidence from a wide range of assignments and tests, and is subject to later review.

38   A small but significant number of secondary schools are giving thought to issues about **equality of opportunity**. A few schools with large proportions of low attaining pupils struggle to find mathematical material which does not require reading skills in advance of pupils' capabilities. Some schools successfully organises sixth form students to work with younger pupils who need help at lunch time. Differences in the examination performance of boys and girls are evident at many schools, and occasionally poor behaviour is noted where boys predominate in lower sets. A few schools monitor these differences regularly, raise the awareness of staff to potential problems and attempt to translate their findings into positive action.

Provision for pupils with special educational needs

39   The quality of provision for pupils with special needs, and the progress made, is very variable. It was judged positively in two-fifths of primary schools and half the middle and secondary schools. In the most successful schools the needs of pupils are identified clearly and a suitable programme of work is devised. Progress is unsatisfactory or poor when the only differentiation is provided by the speed of working through the textbooks or worksheets. Too frequently the reading age

of the material has not adequately been taken into consideration. Schools adopt a variety of organisational arrangements. In Key Stage 1 the help is most frequently given by the class teacher with additional support from a non-teaching assistant. Where special arrangements are made in Key Stage 2, pupils are more likely to be withdrawn for a set time each week. Most frequently in secondary schools help is given to pupils through the setting arrangements; small sets are created for low attainers sometimes at the expense of the more able who are taught in very large classes. Where classes are mixed ability, commonly in Year 7 and Year 8, **low attaining pupils** are either given in-class support or withdrawn. In the latter case a 'bottom set' is essentially created; the effectiveness is dependent on the teachers' knowledge of mathematics and ways of overcoming the learning difficulties of the pupils. Too frequently the in-class support is *ad hoc* and not planned. It is provided by specialist SEN teachers, specialist mathematical teachers, non-specialist teachers, NTAs or other volunteers. It is not possible to generalise about the quality of the organisation in relation to standards – both good and poor provision is associated with each type of organisation.

40 The most successful secondary departments have a policy linked to the overall school policy and links with the SEN department are good. Low attainers have access to the full mathematics curriculum with appropriate emphasis on essential knowledge, skills and understanding; frequently this is based in Years 9–11 on a national scheme which motivates the pupils by providing relatively short-term goals. Pupils make unsatisfactory progress when work is ill-matched to their previous attainment and experiences.

41 It is rare for attention to be given to the needs of **able pupils** other than by placing them in 'top' sets. Some teachers fail to realise that the range of attainment in these classes is wide and so do not provide a sufficiently challenging course. However, a small number of schools are taking initiatives to encourage able pupils; these include the organisation of mathematics clubs and competitions, involvement in national contests, early entry to public examinations, and additional lessons outside of normal school hours as well as greater differentiation in lessons.

Management and administration

42　There is a strong correlation between the **quality of subject management** in the school given by the mathematics co-ordinator or head of department and the standards achieved. The schools which achieve consistently good standards are invariably well-managed and provide clear guidance to teachers. The quality of leadership is of concern in about two-thirds of the primary schools (all of which are large JMI schools) and one-fifth of the secondary schools.

43　Where teachers are given responsibility to co-ordinate mathematics in a primary school, their work is generally more effective when INSET time is used to involve all teachers in discussing and devising curriculum policies. The responsibilities of co-ordinators are often not clearly defined and few have adequate time to develop their role fully. In some schools the co-ordinator does little other than organise documentation and order resources and his or her work has little impact on the curriculum. In a few schools the co-ordinator is responsible for monitoring the implementation of school policies or pupils' progress. In some of the schools where some time has been made available to co-ordinators it is used most effectively to evaluate the quality of the curriculum provided, to provide leadership to teachers who need to extend their experiences, to encourage more consistent approaches throughout the school, and to monitor overall continuity and progression.

44　In secondary schools where management is good, a comprehensive departmental handbook has frequently been produced; curricular practice is regularly reviewed, all teachers of the subject have listed responsibilities, and planning includes detailed INSET proposals. Effective handbooks include a clear statement of philosophy, syllabuses matched to the organisation of teaching groups and guidance about teaching strategies, marking policies and the use of resources.

Resources and their management

Key Stages 1 and 2

45　Few of the schools have teachers who had studied mathematics beyond the minimum professional requirement of their initial training qualifications. While some teachers had gained mathematical qualifications during their teaching careers and some had benefited from GEST-funded 20-day courses, most are still not confident mathematically. The areas in which teachers commonly lack confidence are AT1 (using and applying) and AT5 (handling data), especially aspects of probability and the interpretation of data. INSET opportunities vary enormously. Where teachers are able to attend courses outside school there is usually no systematic sharing of the experience with other colleagues. Some of the schools, particularly in Key Stage 1, make good use of **non-teaching assistants**. The use of assistants is only effective in those classes where they are fully briefed by the teacher.

46   Many schools have a comprehensive range of **practical equipment and materials** to support mathematical activities and to develop understanding. The quality of textbooks is at least adequate in about four-fifths of the schools, although the quality is much lower than those available in the secondary schools. More schools are using material from a variety of sources to supplement their textbook series in order to meet the statutory requirements. Just over a third of the schools, however, fail to make good use of the learning resources they have because of poor planning or lack of awareness of what is available. In many of the schools there is a central store of equipment as well as a supply of materials which are regularly used in each classroom. Generally more thought could be given to the organisation of, and access to, resources for pupils and teachers. The computers which are available in most classrooms are rarely used for the teaching of mathematics. Most pupils learn mathematics in their own classroom and these usually provide adequate environments for a range of mathematical work. It is rare to see a stimulating display of work related to mathematics.

Key Stages 3 and 4 and Sixth Form

47   The majority of departments have a core of **suitably qualified teachers** who teach a range of abilities across Key Stages 3 and 4. A substantial proportion of departments, however, have too few qualified full-time specialists to cover all their work. The difference is usually made up by teachers who have major commitments in other departments or senior management. Unless there is substantial guidance,

weaknesses occur when two or more teachers share the same class. Part-time members do not always participate in full departmental discussion or INSET, and if guidance is weak the curricular experience of their pupils is sometimes very narrow and frequently lacks coherence. Several departments have few teachers with adequate experience or expertise to teach A-level and AS classes. A small number of departments have not recently participated in external INSET; debate and courses arranged by the school itself are often helpful but rarely sufficient to help teachers to keep abreast of national developments.

48   Overall, the **quality of textbook provision** is satisfactory. In the vast majority of schools there is an adequate supply of **calculators** although a few schools cannot provide enough for pupils when they wish to use them as a teaching aid. In an increasing number of schools, sixth form pupils have access to graphical calculators and are making very effective use of them. Three-tenths of the schools fail to make good use of their available resources because of poor planning, limited access or lack of awareness by teachers. The **quality of the resources** gives concern in about a quarter of the schools for two main reasons: the limited range of practical equipment to support learning and of IT facilities. Poor use of IT resources can be the result of lack of access to hardware and difficulty in obtaining appropriate software. In some of the schools, newer computers provide wider access to a good range of standard facilities, such as spreadsheets, while in others the schools have not yet purchased a suitable range of software or provided necessary INSET for staff. Only a relatively small proportion of the schools exploit the potential of computers to assist the teaching and learning of mathematics.

49   Mostly, departments have reasonable **accommodation** which they use appropriately. Approximately two-fifths of schools benefit from good accommodation, for example, where a suite of rooms facilitate collaboration amongst staff in sharing resources, ideas and displays of pupils' work. In the majority of schools staff have their own teaching base but some classes still have lessons in non-specialist rooms with poor access to resources. Where schools have computers based centrally, mathematics departments frequently experience some difficulties with access, particularly where these rooms are used by others as a teaching base.

Inspection issues

Inspection development

50   Inspections carried out under Section 9 of the Education (Schools) Act 1992 began in September 1993. Inspection teams have made a good start in meeting the demanding requirements of the Framework for the Inspection of Schools; they have become more confident as the year has progressed and some early uncertainties have been resolved in many cases. This part of the report draws together some of the key issues for further improving the quality and standard of inspection. Many issues are similar from one subject to another but where there are subject-specific matters these are indicated.

51   Some examples of inspection writing are included. They are not intended to be viewed as models or templates but have been chosen to illustrate how some inspectors have effectively met the Framework requirements.

Evidence gathering

52   Within the time available, inspectors sample the range of work of different year groups, abilities and key stages across the compulsory years of education. A good balance is often achieved. Some aspects, however, tend to receive less emphasis than others. These are work at Key Stage 2 in primary schools, the post-16 work of secondary schools and, where there is setting, middle ability groups.

53   In reaching their judgements, inspectors use evidence from a range of sources. Inspectors need to make clear reference to them in reaching judgements. Much more use could be made of the Supplementary Evidence Form which provides a means of documenting evidence and judgements from sources other than lessons.

Lesson Observation Forms

54   In general, Lesson Observation Forms are completed conscientiously and attention is paid to the evaluation criteria in the Framework. Inspectors could usefully check that on these forms and in other writing, subject-specific character and detail is included wherever possible.

55   In relation to the **content** of lessons, the main theme of lessons is usually indicated. Further details of the mathematical content of lessons and how it relates to the National Curriculum requirements would be helpful in setting the context particularly in relation to the teaching and levels of achievement. Examples of the 'content' section from Lesson Observation Forms follow:

Year 6, upper ability

Introductory work led by the teacher on directed number aimed at Ma2 Levels 4 and 5 with a group of six able pupils. (The remaining 25 pupils continue independently with English.) Textbooks and worksheets were used.

Year 11, lower ability

Second lesson about market research. Designing a questionnaire to conduct a survey (relates to Ma5/4) of best buys at a local super-market. Using a computer database to explore ways of presenting their findings (Ma5/4 and Ma1/4).

56   Responding to the Framework requirements to assess **pupils' achievements** in relation to national expectations and taking account of pupils' abilities has not proved easy. Revised requirements and guidance published in June 1994 will help inspectors in making these distinct judgements. To support judgements it is important that inspectors clearly identify and record the mathematics that pupils know, understand and can do and set achievements in the context of National Curriculum Attainment Targets and Statements of Attainment. Inspectors should cite evidence to inform judgements about pupils' achievements in relation to their abilities. The examples which follow include a number of these features:

Year 6, upper ability

Achievement (age referenced): Pupils competent with mental addition and subtraction of 2 digit numbers (achieving at Level 4 in Ma2) and during the lesson developed their understanding of negative number (Ma2/5) in different contexts, including thermometers and weather charts from newspapers. This level of achievement is slightly above that to be expected of average pupils early in the Autumn term. Grade: 2

Achievement (taking account of pupils' abilities): This level of achievement was appropriate for 5 of the 6 pupils. The sixth grasped the concepts with ease and was not stretched, particularly as recent exercises completed by this boy had been at L6 in Ma3 (series and equations). *Grade: 3*

Year 8, mixed ability (pupils continuing individual work on a range of topics)

Achievement (age referenced): A wide ability range, normally distributed. Considerable variation in achievement but with most achieving at the national expectations for their age. A very able girl was completing an exercise on the use of coordinates in all four quadrants and symmetry properties of polygons (achieving at Level 6 in Ma3 and Ma4). A boy of average ability was working successfully through a routine exercise on percentages (achieving at Level 5 in Ma2). He used a calculator appropriately for most questions but was not able to switch to mental methods for 10%, 25% etc. A low attaining boy constructed a simple bar chart from data tabulated in the text (achieving at Level 3 in Ma5). He was unable to decide what the graph told him. *Grade: 3*

Achievement (taking account of pupils' abilities): Pupils of average ability were achieving appropriately but those at the extremes of the ability range were underachieving. The able girl was repeating work she had mastered and remembered from the previous year; the lower attaining pupil was held back by the nature of the task. Pupils had difficulty explaining the nature of the work and justifying their methods. *Grade: 4*

Year 11, lower ability

Achievement (age referenced): Set 8 from 8 across the year. The teacher's records showed recent work mostly at Level 3 in Ma2 and Ma4. In this lesson most pupils were learning to plan their work methodically and selected graphs and diagrams from the available software to represent their data. This work was at Level 4 but was still well below average attainment at this age. *Grade: 5*

Achievement (taking account of pupils' abilities): The level of achievement demonstrated in this lesson was higher than that *evident in previous work. It was good for these pupils at this stage of the development of the survey. Grade: 2*

57   Inspectors usually cite relevant evidence when judging the **quality of teaching** and evaluation is based on the criteria in the Framework. They need to ensure that the full range of the criteria is used, including teachers' command of the subject and the appropriateness of the work set to the learning needs of the pupils, especially where the range of attainment in any one class is large.

Year 8, mixed ability

The teacher organised the room and the materials appropriately for the style of working. He remained seated at his desk, marked work brought to him and assisted those pupils who came out for help. He was unaware of the progress being made by other pupils. A lack of timely intervention with those pupils resulted in inappropriately matched work, insufficient challenge and poor use of time. Grade: 4

Year 11, lower ability (short exposition on fractions followed by exercises)

Teacher exposition clear but unimaginative. Pupils did not fully understand the explanations. Pace satisfactory. On request, teacher supported pupils but made little proactive intervention. Relationships sound but sometimes familiar. Pupils' work was ticked but comments were very few; no recording of marks/progress. A lot of time spent trying to get pupils to work. The use of a single strategy throughout the lesson was ineffective. Grade: 4

58   Lesson Observation Forms could be more widely used to indicate contributions made by lessons to learning in other areas of the curriculum. They also provide opportunities to signal the impact of contributory factors on achievements and the quality of learning which can be drawn on when compiling the 'contributory factors' sections of the Subject Evidence Form.

Subject Evidence Forms

59 Subject Evidence Forms are usually fully completed. Particular attention is given to aspects of standards of achievement and the quality of learning and teaching, and these sections are usually completed thoughtfully and conscientiously. In most cases it is evident that judgements are based on a range of evidence. Inspectors need to check that this is sufficiently explicit in the sections of the Form. For example, it is important that the basis for evaluating the results in public examinations and National Curriculum Assessments is clear. Completed sections on standards of achievement and the quality of learning follow; they illustrate different styles of presenting evidence.

Standards of Achievement

In 1992 84% of pupils in Y2 achieved L2 in the end of KS1 assessments. This compares with a national average of 78%. No pupils achieved L3 compared with a national average of 9%. No data is available for 1993. Work seen in lessons, material on display and samples of previous work showed that standards of achievement throughout the school are mostly satisfactory and broadly in line with national expectations. Standards of achievement in number and algebra in both KS are at least satisfactory in both key stages, while those in shape and space and handling data are satisfactory but could be raised in KS2. The use and application of mathematics (Ma1) are insufficiently developed.

In KS1 pupils use the language of number including symbols and relationships appropriately. Most pupils can add, subtract, multiply and divide small numbers and many calculate mentally with accuracy. Pupils use calculators appropriately and with confidence for simple number operations. Most use non-standard measures of length by the end of Y1. Most pupils in Y2 can tell the time from analogue clocks, including the use of halves and quarters. Most pupils know the names of regular shapes.

In KS2 most pupils understand place value and can add, subtract, multiply and divide large numbers. Many pupils use standard units of distance, time, weight, rotation and capacity appropriately. In both Key Stages several pupils make sensible estimates. Generally, the development of spatial concepts is less satisfactory particularly in KS2. Many pupils find difficulty creating the nets of 3D solids.

In KS1 and KS2 pupils understand information presented numerically and graphically. Pupils in KS1 create and interpret block graphs. Pupils in KS2 use decimal notation, fractions and graphical representation competently with respect to money and time. They are less confident in using ratio and proportion.

In KS1 pupils handle statistical information in everyday contexts using, for example, data from attendance and dinner registers, but these skills are insufficiently developed in KS2. In KS1 and KS2 pupils recognise pattern to solve elementary problems. Generally, however, standards of problem-solving and investigation are unsatisfactory.

Quality of Learning

- *QoL is satisfactory at both key stages*

- *KS1 children exploit opportunities for play and practical activities well*

- *Children generally persevere with tasks*

- *Co-operative work not well developed*

- *Mathematical language developed well when teacher present with a group*

- *Progress made in majority of lessons – pace slow in some KS2*

- *Insufficient variety of approaches to problem-solving*

- *Independent learning developed inconsistently*

- *A few children able to recognise patterns in work*

- *Appropriate use of a wide range of resources*

- *Mainly accurate and well presented written work – especially KS2*

60   When considering features such as the provision of resources for learning, the organisation, management and procedures in the subject, inspectors should check that any descriptive comments lead to evalua-

tion. The focus of evaluation should be on the impact of these factors on the standards achieved and the quality of learning. For example, clearer evaluation about the effects of the grouping of pupils for mathematics teaching including those with special educational needs or the effectiveness of homework policies would be helpful. Two examples of 'contributory factors' sections follow.

Assessment, Recording and Reporting

Careful consideration has been given to assessment requirements for the end of KS3 and GCSE. As a result pupils are beginning to gain insight into the standards required for particular grades and levels, and staff are able to make increasingly secure assessments. However, the use of marking and other assessments in order to inform future lesson planning is at an early stage of development. Work is marked regularly but pupils receive little helpful feedback. As a result, pupils are not making as much progress as they could.

End of module reviews contain some indication of areas for development, but because of their timing it is not usually practicable for pupils to act upon these.

Learning objectives identified in NC terms enable staff to get reasonably clear pictures of pupil progress against NC. Considerable attention given to assessment of Ma1 with "pupil speak" versions of relevant SoA linked to specific tasks. Staff have periodic discussion on interpretation of Ma1 with appropriate use being made of SEAC publications.

Curriculum Content

All ATs are covered by all pupils in both Key Stages. AT1 is generally addressed as 'bolt on' task-based investigations. There is little overt contribution to cross-curricular themes and dimensions although E&IU is supported, albeit unknowingly, in some work (for example on tax and wages). Use of IT is planned in most years; no IT was observed in use although some pupils spoke enthusiastically about the use of a geometry package.

Pupil groupings are generally sound in terms of ability setting. There is movement between sets both up and down following termly reviews. Lower ability sets are smaller in size than the higher groupings. The setting arrangements are therefore generally effective and in many cases have a positive effect on standards and quality. In some cases, however, the concentration of similar ability pupils has a negative effect on behaviour and consequently on learning (most notably in the middle sets).

Class sizes post-16 are small and are reported by the teacher as having a negative impact on discussion opportunities and peer group pressure and support.

Judgement Recording Statements

61   The proformas of Judgement Recording Statements are usually fully completed. Inspectors need to ensure that the recorded judgements take full account of all available evidence. The purpose and use of Judgement Recording Statements are outlined in Appendix C of Part 3 of the *Handbook for the Inspection of Schools*.

Subject sections in inspection reports

62   Care is obviously taken in writing these sections. They usually give appropriate emphasis to standards of achievement and the quality of learning and teaching; more subject character could, however, be given to these comments. Inspectors should ensure that overall judgements are clear and succinct and draw on all the evidence available. Factors which impact on standards of achievement and quality of learning need to be clearly identified.

63   The following extracts from different reports illustrate writing about standards of achievement, the quality of teaching and contributory factors.

School A

Standards of achievement in mathematics in relation to students' capabilities are generally satisfactory or better in both KS3 and 4. The standards of achievement as judged by the results of the pilot end of Key Stage 3 (1992) tests were in line with national norms. However, the results in the GCSE (1992) for grades A-C were significantly above the national figure and the school's results for

1993 are considerably improved again. The difference between standards achieved by girls and boys, particularly at the higher grades, suggests underachievement by some of the boys.....

School B

.....The quality of teaching is satisfactory or better in only 60% of lessons pre-16; it is good or very good in only 10%. In this small proportion of lessons where teaching is good there is clear planning with realistic goals. Tasks are well matched to pupils' abilities and needs. In the best lessons investigative skills and practical working styles feature prominently. Accurate marking and the use of both motivational and helpful comments increase pupils' confidence and promote progress. In a significant proportion of lessons, where teaching is unsatisfactory or poor, planning is superficial with limited goals. Expectations are unrealistic and often low. Lessons have little pace. Poor achievement, little completed work and inappropriate or poor behaviour are accepted too readily. Teaching is poorest at KS4.

The curriculum at both Key Stages meets statutory requirements. Summative tests are used to assess NC attainment. Little diagnostic assessment is undertaken to inform future planning or promote pupil progress. Records of NC attainment are over-elaborate and consequently inconsistently and sometimes poorly maintained by staff. Both assessment and recording would benefit from further review and development.

School C

.....Three organisational factors affect pupils' standards of achievement. Firstly, 40% of classes at Key Stage 3 have their teaching shared by two teachers; this militates against effective continuity in learning. Secondly, there is a lack of continuity of in-class support from one lesson to the next for large classes of low attaining pupils. Thirdly, the timetabling of lessons at Key Stage 4 results in some classes containing pupils from a wide ability range and with different teaching needs which are difficult to fully meet.....

64   In writing to the Framework requirements, inspectors need to check that a comment is included on compliance with statutory

requirements and that key issues for action in mathematics are included. These are helpful to schools in their action planning.

The interpretation of subject performance data

National Curriculum assessments

65   Key Stage 1 data is attached as Annex A. Just over 80% of pupils achieved Level 2 or above in mathematics (Teacher Assessment) and in the Number Attainment Target. Other significant features are:

- 18% achieved Level 3 in Number, compared with 12% in each of Algebra and Shape and Space, 11% in Using and Applying and 9% in Handling Data.

- The same proportion of boys and girls achieved Level 3 although a slightly higher proportion of girls than boys – 83% rather than 79% – reached at least Level 3.

66   Key Stage 3 data is attached as Annex B. Significant features are:

- 65% of pupils achieved at Level 5 or above on Teacher Assessment and 60% on the Tests.

- Girls slightly outperformed boys, with 63% at or above Level 5 on the Tests compared with 60% of boys. Comparative figures for Teacher Assessment are 67% and 60% respectively. In both Teacher Assessment and Test, 1% more boys than girls achieved Level 8.

GCSE

67   Data is attached as Annex C. Key features of the results in 1994 are given below.

- There was an improvement on the 1993 figures, which are given in brackets below.

- 92% (89%) of the cohort registered in maintained schools (excluding special schools) entered for mathematics; 91% of boys and 92% of girls.

- 87% (84%) of the cohort achieved grades A*-G; 86% of boys and 88% of girls.

- The average point score of those entered was 4.09 (3.82) compared with 4.59 for English Language and 4.41 for double award science.

- 42% of the entry, 39% of the cohort, achieved grades A*-C.

While girls generally perform better than boys at GCSE in terms of the proportion gaining A*-C, boys' best result relative to girls is in mathematics where they were ahead by 0.7%. The average point score for boys (4.11) is marginally better than that for girls (4.06). When considering the cohort 38.8% of boys and 38.7% of girls achieved grades A*-C.

It would be helpful if inspectors noted any significant imbalance in achievements of boys and girls in comparison with national figures for boys and girls and pursued any reasons for these.

GCE A/AS

68   Data is attached as Annex D. The results in the table may not include all the results of those taking modular examinations.

The most significant features of the results are:

- Entries for A-level mathematics appear to be down by about 5% from 1993 while the total number of entries for A-level is down by 0.4% against a 4.0% decline in the number of 17-year-olds in the population.

- The ratio of boys to girls entering for the examination remains about 2:1.

- There has been an improvement in the A-Level results in common with other subjects. For example, over 37% of the entry achieved a grade A or B compared with almost 36% in 1993 and 30% in 1992. Over recent years the proportion obtaining 'U' has decreased, suggesting that students are receiving more appropriate advice about subjects.

- A worryingly high proportion of AS students (over 30%) does not achieve success in this examination and the average point score (1.60) is low.

It would be helpful if inspectors commented on the policies for choice of syllabuses and entry. The use and effect of new A/AS syllabuses (SMP, MEI etc) and/or modular courses is of particular interest. Comments about changing entry patterns would be helpful.

Annex A

National Curriculum assessment at Key Stage 1

Key Stage 1 Attainments in mathematics

The results are based on responses from about 52% of maintained schools and 10% of independent schools. DFE believes it is likely that the results for these schools may overstate slightly the national estimates that would have been obtained if all schools had participated this year.

Percentage of pupils achieving Level 2 or above

	TA	Test
English	79	-
Reading	80	80
Writing	70	67
Spelling	73	71
Handwriting	81	80
Mathematics	81	-
Number	82	81
Science	85	-

Approximate percentages of pupils achieving each level in mathematics

	TA Mathematics	Test Number	TA Number
4	0	0	0
3	11	22	18
2	70	59	64
1	18	16	16
D and W	1	3	1

Percentage of pupils achieving at least Level 2 (TA) in each AT

Using and Applying   78
Number   82
Algebra   82
Shape and Space   77
Handling Data   79

Eighteen percent achieved Level 3 in Number, compared with 12% in each of Algebra and Shape and Space, 11% in Using and Applying and 9% in Handling Data.

The same proportion of boys and girls achieved Level 3 although a slightly higher proportion of girls than boys – 83% rather than 79% – reached at least Level 3.

Annex B

National Curriculum assessment at Key Stage 3

Key Stage 3 Attainments in mathematics

Results are based on responses from about 22% of maintained schools and 9% of independent schools. DFE believes the sample is skewed upwards; the available figures from these schools are, therefore, likely to overstate the national estimates towards the higher end of the ability range. Comparisons with 1992 results are not possible because 1992 was a pilot year.

Percentage of pupils achieving Level 5 or above

	TA	Test
English	63	58
Mathematics	65	60
Science	65	64

Approximate percentages of pupils achieving each level in mathematics

	TA %	Test %
9 and 10	<0.5	<0.5
8	3	2
7	13	10
6	25	26
5	24	22
4	20	19
≤3	15	15

Girls slightly out-performed boys, with 63% at or above Level 5 on the Tests compared with 60% of boys. Comparative figures for TA are 67% and 60% respectively. In both Teacher Assessment and Test, 1% more boys than girls achieved Level 8.

Annex C

GCSE Mathematics results for 15 year olds[1] for 1994

Type of School		Number of 15[1] year old pupils entered	Percentages achieving grades									1994			1993	
			A*	A	B	C	D	E	F	G	U	Average points score[2]	% A*–C grades	% A*–G grades	Average points score[2]	% A–C grades
Maintained	All pupils	437005	1.4	6.3	14.8	19.8	15.4	16.1	13.7	7.4	2.2	4.09	42.3	94.8	3.82	41.8
	Boys	219767	1.7	6.8	14.8	19.3	15.0	15.7	13.6	7.6	2.2	4.11	42.7	94.6	3.83	42.0
	Girls	217238	1.1	5.8	14.7	20.3	15.8	16.5	13.7	7.2	2.2	4.06	42.0	95.1	3.82	41.7
Comprehensive		405664	1.3	5.6	14.0	19.9	15.7	16.5	14.1	7.6	2.3	4.02	40.8	94.7	3.77	40.5
Selective		16424	6.7	29.0	38.2	19.7	3.8	1.3	0.6	0.3	0.0	6.07	93.6	99.7	5.94	93.0
Modern		14917	0.2	1.3	8.5	16.8	18.8	21.7	17.4	9.2	2.8	3.51	26.8	93.9	3.20	25.0
All subjects			2.1	8.4	16.4	20.5	18.9	14.5	10.2	4.5	1.5	4.40	47.4	95.5	4.12	46.1

1 Aged 15 on 31/8/93
2 Calculated on basis A*=8, A=7, B=6, C=5, D=4, E=3, F=2, G=1

Annex D

GCE A-Level Mathematics results 1994

Type of School		Number of candidates	Percentages achieving grades							% A–B grades	% A–E grades	Average points score[1]	1993 % A–B grades	1993 % A–E grades	1992 % A–B grades	1992 % A–E grades
			A	B	C	D	E	N	U							
Maintained	All pupils	27006	20.8	16.5	16.4	14.6	11.7	8.2	7.4	37.3	80.0	5.20	35.9	78.4	29.5	73.4
	Boys	17682	22.2	15.8	15.4	14.3	11.8	8.5	7.6	37.9	79.4	5.21	36.5	78.0	29.8	72.8
	Girls	9324	18.3	17.8	18.4	15.1	11.6	7.6	7.1	36.1	81.1	5.19	34.7	79.3	28.9	74.6
Comprehensive		21567	19.3	15.7	16.4	14.9	12.4	8.9	8.2	35.1	78.8	5.02	33.6	76.9	28.0	72.1
Selective		5290	27.3	19.5	16.4	13.3	8.8	5.5	4.0	46.8	85.2	5.98	45.6	85.1	38.3	81.7
Modern		149	6.0	15.4	14.1	15.4	18.8	7.4	11.4	21.5	69.8	3.68	15.6	53.9	21.5	62.3
All subjects			13.1	16.2	18.5	18.9	15.2	9.4	7.5	29.3	81.9	4.78	28.0	79.7	26.4	78.6

1 Calculated on basis A = 10, B = 8, C = 6, D = 4, E = 2.
The results in the table may not include all the results of those taking modular examinations.

GCE AS Mathematics results 1994

Type of School		Number of candidates	1994							% A–B grades	% A–E grades	Average points score[1]	1993		1992	
			Percentages achieving grades										% A–B grades	% A–E grades	% A–B grades	% A–E grades
			A	B	C	D	E	N	U							
Maintained	All pupils	4072	8.1	8.8	13.0	15.0	15.6	11.6	20.9	16.9	60.4	1.60	18.2	57.4	18.9	61.5
	Boys	2616	8.2	8.2	11.5	14.5	15.6	12.3	22.2	16.4	58.0	1.53	18.3	56.4	18.8	60.7
	Girls	1456	8.0	9.8	15.7	15.8	15.5	10.2	18.7	17.7	64.8	1.73	18.1	59.3	19.1	63.3
All subjects			7.1	10.2	14.8	17.9	18.2	12.9	15.1	17.3	68.2	1.75	17.0	65.5	16.6	65.4

1 Calculated on basis A = 5, B = 4, C = 3, D = 2, E = 1.
The results in the table may not include all the results of those taking modular examinations.

Printed in the United Kingdom for HMSO
Dd300284 4/95 C130 G3397 10170